What Season Is It?

by Caroline Hutchinson

Spring is here. The flowers are blooming. What can we do in the spring?

We can pick flowers

in the spring.

Here comes summer.
The grass is green.
What can we do
in the summer?

We can have a picnic
in the summer.

Fall is here. The apples are on the trees. What can we do in the fall?

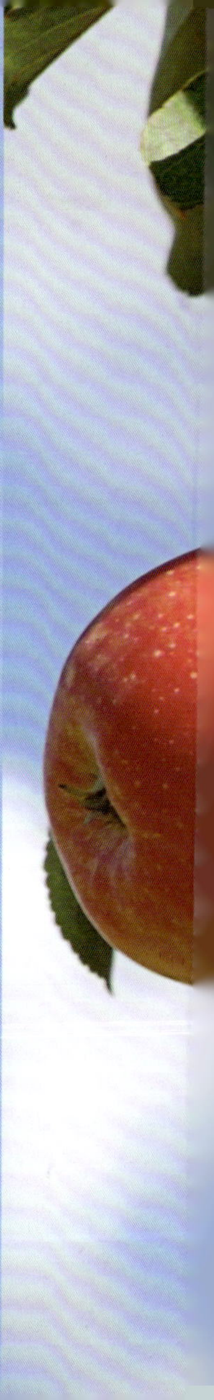

We can pick apples
in the fall.

Here comes winter. It is snowing.
What can we do in the winter?

We can make a snowman!